BACKYARD BUGS

I SEE LADYBUGS

by Genevieve Nilsen

TABLE OF CONTENTS

tadpole
books

I SEE LADYBUGS

I see a bug.

red

It is red.

spot

It has spots.

leg

It has legs.

wing

It has wings.

It eats.

It is a ladybug!

WORDS TO KNOW

eats

ladybug

legs

red

spots

wings

INDEX